Percent: Grade 7
A complete workbook with lessons and problems

By Maria Miller

Copyright 2017 Maria Miller.
ISBN 978-1533160621

EDITION 5/2018

All rights reserved. No part of this workbook may be reproduced or transmitted in any form or by any means, electronic or mechanical, or by any information storage and retrieval system, without permission in writing from the author.

Copying permission: Permission IS granted for the teacher to reproduce this material to be used with students, not for commercial resale, by virtue of the purchase of this workbook. In other words, the teacher MAY make copies of the pages to be used with students.

Contents

Preface	5
Introduction	7
Helpful Resources on the Internet	8
Review: Percent	13
Solving Basic Percentage Problems	16
Percent Equations	19
Circle Graphs	24
Percentage of Change	26
Percentage of Change: Applications	29
Comparing Values Using Percentages	33
Simple Interest	37
Review	43
Answers	45
Appendix: Common Core Alignment	57

Preface

Hello! I am Maria Miller, the author of this math book. I love math, and I also love teaching. I hope that I can help you to love math also!

I was born in Finland, where I also grew up and received all of my education, including a Master's degree in mathematics. After I left Finland, I started tutoring some home-schooled children in mathematics. That was what sparked me to start writing math books in 2002, and I have kept on going ever since.

In my spare time, I enjoy swimming, bicycling, playing the piano, reading, and helping out with Inspire4.com website. You can learn more about me and about my other books at the website MathMammoth.com.

This book, along with all of my books, focuses on the conceptual side of math... also called the "why" of math. It is a part of a series of workbooks that covers all math concepts and topics for grades 1-7. Each book contains both instruction and exercises, so is actually better termed *worktext* (a textbook and workbook combined).

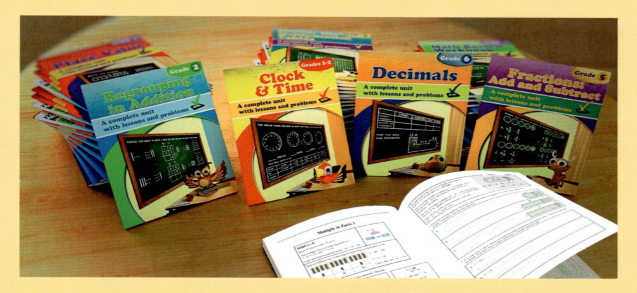

My lower level books (approximately grades 1-5) explain a lot of mental math strategies, which help build number sense — proven in studies to predict a student's further success in algebra.

All of the books employ visual models and exercises based on visual models, which, again, help you comprehend the "why" of math. The "how" of math, or procedures and algorithms, are not forgotten either. In these books, you will find plenty of varying exercises which will help you look at the ideas of math from several different angles.

I hope you will enjoy learning math with me!

Introduction

Percent: Grade 7 Workbook begins by reviewing the concept of percent as "per hundred" or as hundredth parts and how to convert between fractions, decimals, and percentages. The second lesson in the workbook, *Solving Basic Percentage Problems*, is intended for review of sixth grade topics, focusing on finding a known percentage of a number (such as 21% of 56) or finding a percentage when you know the part and the total.

We take a little different perspective of these concepts in the lesson *Percent Equations*. Students write simple equations for situations where a price increases or decreases (discounts). This lesson also explains what a percent proportion is. Personally, I prefer *not* to use percent proportion but to write the percentage as a decimal and then write an equation. I feel that approach adapts better to solving complex problems than using percent proportion.

Here is a quick example to show the difference between the two methods. Let's say an item is discounted by 22% and it now costs $28. Then, the new price is 78% of the original. If we let p be the price of the item before the discount, we can write the percent proportion $\$28/p = 78/100$ and solve for p. If, we write the percentage 78% as the decimal 0.78, we get the equation $0.78p = \$28$. Personally, I consider percent proportion to be an optional topic, and the reason I have included it here is to make this curriculum fully meet the Common Core Standards for seventh grade.

The lesson *Circle Graphs* provides students a break from new concepts and allows them to apply the concept of percent in a somewhat familiar context. Next, we delve into the percentage of change. Students sometimes view the percentage of change as a totally different concept as compared to other percentage topics, but it is not that at all. To calculate the percentage of change, we still use the fundamental idea of *percentage = part/total*, only this time, the "part" is how much the quantity in question changes (the difference) and the "total" is the original quantity.

Tying in with percentage of change, students also learn to compare values using percentages, such as how many percent more or less one thing is than another. Once again, this is not really a new concept but is based on the familiar formula *percentage = part/total*. The percentage difference (or relative difference) is the fraction (*actual difference*)/(*reference value*).

Simple Interest is a lesson on the important topic of interest, using as a context both loans and savings accounts. Students learn to use the formula $I = prt$ in a great variety of problems and situations.

The text concludes with a review lesson of all of the concepts taught in the workbook.

I wish you success in teaching math!

Maria Miller, the author

Helpful Resources on the Internet

Use these free online games and resources to supplement the "bookwork" as you see fit.

Percent videos by Maria
Videos on percent-related topics that match the lessons in this workbook.
http://www.mathmammoth.com/videos/prealgebra/pre-algebra-videos.php#percent

Percent worksheets
Create an unlimited number of free customizable percent worksheets to print.
http://www.homeschoolmath.net/worksheets/percent-decimal.php
http://www.homeschoolmath.net/worksheets/percent-of-number.php
http://www.homeschoolmath.net/worksheets/percentages-words.php

PERCENTAGES, FRACTIONS, AND DECIMALS

Mission: Magnetite
Hacker tries to drop magnetite on Motherboard. To unlock a code to stop him, match up percentages, fractions, and images showing fractional parts in five different sets of items.
http://pbskids.org/cyberchase/media/games/percent/index.html

Decention Game
Build teams of three players by matching fractions, decimals, and percentages with the same value.
https://www.mathplayground.com/Decention/index.html

Fractions and Percent Matching Game
A simple matching game: match fractions and percentages.
http://www.mathplayground.com/matching_fraction_percent.html

What percentage is shaded?
Practice guessing what percentage of the pie chart has been shaded yellow in this interactive activity.
http://www.interactivestuff.org/sums4fun/pietest.html

Pie Chart and Questions
First, read a short illustrated lesson about pie charts. Then, click on the questions at the bottom of the page to practice.
https://www.mathsisfun.com/data/pie-charts.html

Flower Power
Grow flowers and harvest them to make money in this addictive order-'em-up game. Practice ordering decimals, fractions, and percentages.
https://www.mangahigh.com/en/games/flowerpower

Matching Fractions, Decimals, and Percentages
A simple matching memory game.
http://nrich.maths.org/1249

Sophie's Dominoes
Order dominoes that contain either numbers or a percentage of a number (such as 15% of 300).
http://www.bsquaredfutures.com/pluginfile.php/212/mod_resource/content/1/doms.swf

Percent Goodies: Fraction-Decimal-Percent Conversions
Practice conversions between fractions, decimals and percents. There are three levels of difficulty and instant scoring for each. Note that fractions must be written in lowest terms.
http://www.mathgoodies.com/games/conversions/

BASIC PERCENTAGE CALCULATIONS

Penguin Waiter
A simple game where you calculate the correct tip to leave the waiter (levels "easy" and "medium"), the percentage that the given tip is (level "hard"), or the original bill (level "Super Brain").
https://www.funbrain.com/games/penguin-waiter

Percent Jeopardy
An interactive jeopardy game where the questions have to do with a percentage of a quantity.
http://www.quia.com/cb/42534.html

Matching Percentage of a Number
Match cards that ask for a percentage of a number (such as 75% of 40) with the values. The game is fairly easy and can be completed using mental math.
http://www.sheppardsoftware.com/mathgames/percentage/MatchingPercentNumber.htm

Discount Doors
Calculate the price after the discount.
http://www.bsquaredfutures.com/pluginfile.php/214/mod_resource/content/1/doors.swf

The Percentage Game
This is a printable board game for 2-3 players that practices questions such as
"20 percent of ___ is 18" or "___ is 40 percent of 45".
http://nzmaths.co.nz/resource/percentage-game

A Conceptual Model for Solving Percent Problems
A lesson plan that uses a 10 x 10 grid to explain the basic concept of percent and to solve various types of percentage problems. The lesson includes seven different word problems to solve. Please note their solutions are included on the same page.
http://illuminations.nctm.org/LessonDetail.aspx?id=L249

PERCENT OF CHANGE

Percent of Change Matching
Match five flashcards with given increases or decreases (such as "25 is decreased to 18") with five percentages of increase/decrease.
https://www.studystack.com/matching-182854

Percent Shopping
Choose toys to purchase. In level 1, you find the sale price when the original price and percent discount are known. In level 2, you find the percent discount (percent of change) when the original price and the sale price are known.
http://www.mathplayground.com/percent_shopping.html

Rags to Riches: Percent Increase or Decrease
Answer simple questions about percent increase or decrease and see if you can win the grand prize in the game.
http://www.quia.com/rr/230204.html

Percentage Change 1
A self-check quiz with 10 questions about percentage change. The link below goes to level 1 quiz, and at the bottom of that page you will find links to level 2, 3, 4, 5 and 6 quizzes.
http://www.transum.org/software/SW/Starter_of_the_day/Students/PercentageChange.asp

Percent of Change Quiz
Practice determining the percent of change in this interactive multiple-choice quiz.
http://www.phschool.com/webcodes10/index.cfm?wcprefix=bja&wcsuffix=0607&area=view

Percentage increase and decrease 4 in a line
The web page provides a game board to print. Players take turns picking a number from the left column, and increase or decrease it by a percentage from the right column. They cover the answer on the grid with a counter. The first player to get four counters in a line wins.
https://www.tes.co.uk/teaching-resource/percentage-increase-and-decrease-4-in-a-line-6256320

Percentage of Increase Exercises
Find the percentage increase given the original and final values in this self-check quiz about percentage change.
http://www.transum.org/software/SW/Starter_of_the_day/Students/PercentageChange.asp?Level=3

Percentage of Decrease Exercises
Find the percentage decrease given the original and final values in this self-check quiz about percentage change.
http://www.transum.org/software/SW/Starter_of_the_day/Students/PercentageChange.asp?Level=4

Treasure Hunt - Percentage Increase and Decrease
The clues of this treasure hunt are printable percentage increase/decrease questions.
https://www.tes.co.uk/teaching-resource/treasure-hunt--percentage-increase-and-decrease-6113809

Percent Change Practice
Interactive flash cards with simple questions about percentage of change with three difficulty levels.
http://www.thegreatmartinicompany.com/percent-percentage/percent-change.html

Percentage Increase and Decrease
Multiple-choice questions about percentage of change to be solved without a calculator (mental math).
https://www.mangahigh.com/en/maths_games/number/percentages/percentage_increase_and_decrease_-_no_calculator

Percent of Change Jeopardy
This is an online jeopardy game that provides you the game board, questions for percent increase, percent decrease, sales tax, discounts, and markups, the answers, and a scoreboard where you can enter the teams' points. However, it doesn't have a place to enter answers and requires someone to supervise the play and the teams' answers.
http://www.superteachertools.us/jeopardyx/jeopardy-review-game.php?gamefile=2245685

Percentage Difference
A short lesson about percentage difference, followed by practice questions at the bottom of the page.
http://www.mathsisfun.com/percentage-difference.html

INTEREST

Interest (An Introduction)
Read an introduction to interest, and then click on the questions at the bottom of the page to practice.
https://www.mathsisfun.com/money/interest.html

Quiz: Simple Interest
A quiz with five questions that ask for the interest earned, final balance, interest rate, or the principal.
https://www.cliffsnotes.com/study-guides/algebra/algebra-ii/word-problems/quiz-simple-interest

Simple Interest
Another quiz where you need to find the principal, the amount of time, interest earned, or the final amount in an account earning interest. Four out of nine questions have to do with terminology and the rest are math problems.
http://www.proprofs.com/quiz-school/story.php?title=simple-interest

Simple Interest Game
Answer questions about simple interest by clicking on the correct denominations in the cash register.
http://www.math-play.com/Simple-Interest/Simple-Interest.html

Simple Interest Practice Problems
Practice using the formula for simple interest in this interactive online activity.
http://www.transum.org/Maths/Activity/Interest/

Calculating simple interest
This page includes several video tutorials plus a short three-question quiz on simple interest.
https://www.sophia.org/concepts/calculating-simple-interest

Compound interest
A simple introduction to compound interest with many examples.
http://www.mathsisfun.com/money/compound-interest.html

GENERAL

Percents Quiz
Review basic percent calculations with this short multiple-choice quiz.
http://www.phschool.com/webcodes10/index.cfm?wcprefix=bja&wcsuffix=0605&area=view

Percents Quiz
Practice determining what percentage one number is of another in this interactive online quiz.
https://www.thatquiz.org/tq-3/?-j1c0-l7-mpnv600-p0

Math At the Mall
Practice percentages while shopping at a virtual mall. Find the percentage of discount and the sales price, calculate the interest earned at the bank, compare health memberships at the gym and figure out how much to tip your waiter at the Happy Hamburger.
http://www.mathplayground.com/mathatthemall2.html

Percent Word Problems
Practice solving word problems involving percents in this interactive online activity.
https://www.khanacademy.org/math/algebra-basics/basic-alg-foundations/alg-basics-decimals/e/percentage_word_problems_1

Percents Quiz 2
Practice answering questions about percent in this multiple-choice online quiz.
http://www.phschool.com/webcodes10/index.cfm?wcprefix=bja&wcsuffix=0606&area=view

Percentages Multiple-Choice Test
Test your understanding of percentages with this self-check multiple choice quiz.
http://www.transum.org/Software/Pentransum/Topic_Test.asp?ID_Topic=28

Solving Problems with Percent
Reinforce your skills with these interactive word problems.
http://www.buzzmath.com/Docs#CC07E253

Review: Percent

Percent (or **per cent**) means *per hundred* or "divided by a hundred." (The word "cent" means one hundred.) So, simply put, percent means a hundredth part.

To convert percentages into fractions, simply read the "per cent" as "per 100." Thinking of hundredths, you can also easily write them as decimals.

Therefore, 8% = 8 per cent = 8 per 100 = 8/100 = 0.08.
Similarly, 167% = 167 per 100 = 167/100 = 1.67.

$$\frac{5}{100} \begin{matrix} \text{five} \\ \text{per} \\ \text{cent} \end{matrix} = 5\%$$

1. Write as percentages, fractions, and decimals.

| a. 52% = ___ = ___ | b. ___% = ___ = 0.07 | c. ___% = $\frac{59}{100}$ = ___ |
| d. 109% = ___ = ___ | e. ___% = $\frac{382}{100}$ = ___ | f. 200% = ___ = ___ |

A decimal number with two decimal digits is in hundredths, so it can easily be written as a percentage. For example, 0.56 = 56%. But even if we have 3 or more decimals, we can still convert into percent.

Example 1. The number 0.564 is 564 thousandths. As a percentage, 0.564 = 56.4%. Compare this to 0.56 = 56%. The decimal digit "4" that follows the digits "56" is in the thousandths place, so it becomes 4 tenths of a percent (56.**4**%).

$$0.091 = 9.1\% = \frac{91}{1000}$$

$$0.387 = 38.7\% = \frac{387}{1000}$$

This is how to convert percentages with even more decimal digits:

| decimal | percentage | | decimal | percentage |
| 0.38429 | = 38.429% | | 3.0281930 | = 302.81930% |

Think of it this way. Since 0.38 = 38%, any decimal digits that we have beyond 0.38 (the digits 429) simply become decimal digits for the percentage. In effect, we move the decimal point two places to the right.

2. Write as percentages, fractions, and decimals.

a. 28.2% = ___ = ___	b. 6.7% = ___ = ___	c. ___% = ___ = 0.891
d. 0.9% = ___ = ___	e. ___% = $\frac{1039}{10000}$ = ___	f. ___% = $\frac{3409}{1000}$ = ___
g. 45.39% = 0.___	h. 2.391% = 0.___	h. ___% = 0.942834

13

Writing fractions as percentages

Sometimes you can easily convert a fraction to an equivalent fraction with a denominator of 100. After that it's easy to write it as a decimal and as a percentage.

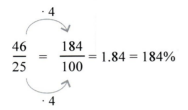

$$\frac{46}{25} = \frac{184}{100} = 1.84 = 184\%$$

For most fractions, you need to divide in order to convert them into decimals first and then into percentages.

Simply treat the fraction line as a division symbol and divide. You will get a decimal. Then write the decimal as a percentage.

Example 1. $\frac{8}{9} = 0.888... \approx 0.889 = 88.9\%$

```
   0.8 8 8 8
9)8.0 0 0 0
  -7 2
     8 0
    -7 2
       8 0
      -7 2
         8 0
        -7 2
           8
```

3. Write the fractions as percentages.

a. $\frac{8}{25} = \frac{}{100} = $ _____ %

b. $\frac{142}{200} = \frac{}{100} = $ _____ %

c. $\frac{24}{20} = \frac{}{100} = $ _____ %

4. Write as percentages. Use long division. Round your answers to the nearest tenth of a percent.

a. 11/8

b. 11/24

5. Use a calculator to convert the fractions into decimals. Round the decimals to four decimal digits. Then write the decimals as percentages.

a. $\frac{2}{3} \approx$ _0.6667_ = _66.67_ %

b. $\frac{6}{7} \approx$ _____ = _____ %

c. $\frac{17}{23} \approx$ _____ = _____ %

d. $\frac{52}{98} \approx$ _____ = _____ %

If you are asked the percentage:	**Example 3.** What percentage is 14 km of 75 km?
Asking what percent(age) is essentially the same as asking "what part" or "what fraction."	1. Write the fraction *part/total*: it is 14/75.
1. Simply write the fraction $\frac{part}{total}$.	2. Then use a calculator and write it as a decimal. $14/75 = 0.18\overline{6}$. Now write the decimal as a percentage: $0.18\overline{6} = 18.\overline{6}\%$.
2. Convert the fraction into a decimal, and then into a percentage.	Normally, we round the result and say that 14 km is about 19% of 75 km.

6. The circle graph shows the areas of the world's five oceans in square kilometers. The total area of these oceans is 335,258,000 km². To the nearest tenth of a percent, find how many percent each ocean is of the total area of the oceans.

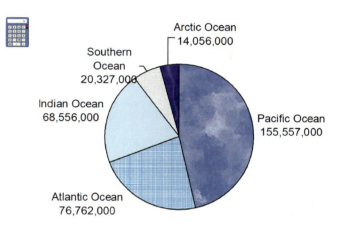

Ocean	Percentage of total area
Pacific Ocean	
Atlantic Ocean	
Indian Ocean	
Southern Ocean	
Arctic Ocean	

7. The Carters live on a rectangular piece of land that measures 40 m × 35 m. The Joneses live on a rectangular piece of land that measures 42 m × 39 m. To the nearest tenth of a percent, find what percentage the area of the Carters' land is of the area of the Joneses' land.

8. Harry has two roosters, named Captain and Chief. The weight of Captain is 7/5 of the weight of Chief.

   a. Write the second sentence above using a percentage instead of a fraction.

   b. If Chief weighs 6 lb, how much does Captain weigh?

Puzzle Corner

What percentage of each figure is colored?

a.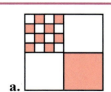

b.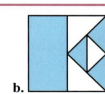

Solving Basic Percentage Problems

If the percentage is known and the total is known: (What is x% of y?) 1. Write the percentage as a decimal. 2. Multiply that decimal by the total. Or use mental math tricks for finding 1%, 10%, 20%, 30%, 25%, 50%, 75%, *etc.* of a number.	If you are asked the percentage: Asking "what percentage" is essentially the same as asking "what part" or "what fraction." 1. Write the fraction $\frac{part}{total}$. 2. Convert that fraction into a decimal, and then into a percentage.

Example 1. A shirt that cost $34 was discounted by $4. What is the percentage it was discounted?

We write the fraction $\frac{part}{total}$ and get $\frac{\$4}{\$34}$ = 2/17 = 0.1176471 ≈ 11.8%.

Example 2. Find 59.2% of $2,600.

Write 59.2% as 0.592, and translate the word "of" into multiplication. We get 0.592 · $2,600 = $1,539.20.

Example 3. A meal that cost $14 was increased by 20%. What is the new price?

Because the price is increasing, the new price isn't 20% but 100% + 20% = 120% of the original one. So we can rewrite the previous sentence as: The new price = 1.2 · $14 = $16.80

In this case, since the numbers are easy, we could also use mental math. Ten percent of $14 is $1.40, so the 20% increase is $2.80. The sum of the original price and the increase gives the new price: $14 + $2.80 = $16.80.

1. Change the percentages into decimals.

a. 107%	**b.** 16.67%	**c.** 4.5%

2. Calculate the new, increased prices. Write the percentages as decimals and use multiplication.

   a. Laptop: Original price $249.90, increase 6%.

   New price = _____ · $249.90 = _____

   b. Biology textbook: Original price $82.40, increase 2.5%.

   New price = _____ · $82.40 = _____

   c. Lunch buffet: Original price $18.50, increase 11.2%.

   New price = _____ · $18.50 = _____

3. Julia paid $325.08 of her $1,890 paycheck in taxes. What percentage of her paycheck did she pay in taxes?

4. Fill in Gloria's solution to the following problem.

Roller blades originally cost $45.50, but now they are discounted by 13%. What is the new price?

Since 13% of the price is removed, _____ % of the price is left. I write that percentage as a decimal and multiply the original price by it: _____ · $45.50 = _____ .

5. Calculate the discounted prices. Write the percentages as decimals and use multiplication.

   a. Subscription to a magazine: Original price $78, discount 38%.

   new price = _____ · $78 = _____

   b. Swimming goggles: Original price $14.95, discount 22.5%.

   new price = _____ · $14.95 = _____

6. The two rectangles are similar.

   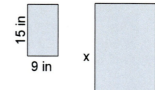

   a. In what ratio are the corresponding sides of the rectangles?

   b. In what ratio are their areas?

   c. Calculate what percentage the area of the smaller rectangle is of the area of the larger rectangle.

Sales Tax and Mental Math

A ticket to a circus costs $25. The sales tax is 6%. What is the final price you have to pay?

The sales tax is always added to the **base price** (the price without tax). We simply calculate 6% of $25 and add that amount to $25.

Calculate in your head: 1% of $25 is $0.25. Six times that is $1.50. So the final price is $26.50.

---

7. Find the final price when the base price and sales tax rate are given. This is a mental math workout, so don't use a calculator!

a. Bicycle: $100; 7% sales tax.	**b.** Fridge: $400; 6% sales tax.	**c.** Haircut: $50; 3% sales tax.
Tax to add: $_____	Tax to add: $_____	Tax to add: $_____
Price after tax: $_____	Price after tax: $_____	Price after tax: $_____

8. Here's another mental math workout! Once again, don't use a calculator. The sales tax is 5%. For each set of items find the price with tax. *Hint: To find 5% of a number, first find 10% and take half of that.*

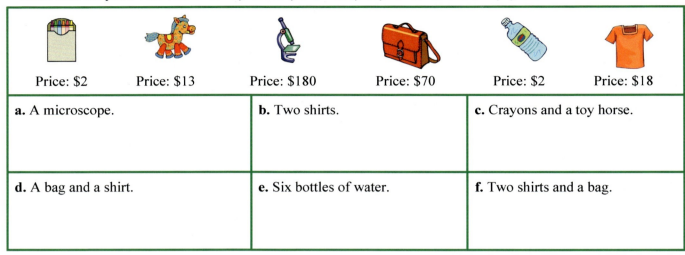

| Price: $2 | Price: $13 | Price: $180 | Price: $70 | Price: $2 | Price: $18 |

a. A microscope.	**b.** Two shirts.	**c.** Crayons and a toy horse.
d. A bag and a shirt.	**e.** Six bottles of water.	**f.** Two shirts and a bag.

9. Find the final price of a music CD with a base price of $11.50 when the sales tax is 6.7%.

10. Jeremy gets a 37.5% discount on a vacation package that normally costs $850. Find what Jeremy will pay for the vacation package.

11. Patrick bought 5,000 m² of land for a base price of $200,000. A 1.2% sales tax was added.

   a. Find the total price Patrick paid.

   b. Later, Patrick decides to sell 2,000 m² of the land to a neighbor. What should Patrick charge his neighbor in order to break even on what he paid for the part he's selling?

12. **a.** The base price of a music CD is $12.50. It is first discounted by 20% and then a 7% sales tax is added. What is the final price of the CD?

   b. The base price of a pair of jeans is $55.97. They are first discounted by 40% and then a 5% sales tax is added. What is the final price of the jeans?

13. Roger compared the unit prices of four different kinds of pasta. One kind cost $2, another $1.50, another $2.20, and another $1.70.

   a. Find the average price of the four types of pasta.

   b. If each type of pasta were discounted by 10%, then what would the average price be?

Percent Equations

Example 1. A handbag has been discounted by 23%, so now it costs $43.81. What was its original price?

Solution with an equation:

Let p be the original price. A discount of 23% means that 23% of the price (or $0.23p$) is taken away from the price (p). As an expression, the discounted price is therefore $p - 0.23p$, which simplifies to $0.77p$.

You can also reason that, after the discount, 77% of the price is left, so the discounted price is $0.77p$.

Since the discounted price is $43.81, the equation to solve is

$$0.77 \cdot p = \$43.81$$

To solve the equation, simply divide both sides by 0.77:

$$0.77p = \$43.81 \quad | \div 0.77$$
$$p \approx \$56.90$$

Check:

$$0.77 \cdot \$56.90 \stackrel{?}{=} \$43.81$$
$$43.813 \approx \$43.81 \checkmark$$

Solution with logical (proportional) reasoning:

Again, we start out by reasoning that 77% of the price is left. In the chart on the right, the long lines mean "corresponds to" (not "equals").

77% —— $43.81

1% —— $43.81/77

100% —— $43.81/77 · 100

100% —— $56.90

1. Write an expression for the final price using a decimal for the percentage.

   a. Headphones: price $12, discounted by 24%. New price = _____

   b. A bag of dog food: price p, discounted by 11%. New price = _____

   c. Pizza sauce: price x, discounted by 17%. New price = _____

   d. Sunglasses: price s, price increased by 6%. New price = _____

2. A computer is discounted by 25%, and now it costs $576. Let p be its price before the discount. Select the equation that matches the statement above and solve it.

   $p + 0.25p = 576$

   $p - 0.25p = 576$

   $0.25p = 576$

3. The rent was increased by 5% and is now $215.25. What was the rent before the increase? Write an equation for this situation and solve it.

4. A tablet is discounted by 30%. Matthew bought two of them, and he paid $98. Find the price of the tablet before the discount (p).

   a. Find the equation on the right that matches the problem.

   b. Solve it.

   $2(p - 30) = 98$

   $2p - 30 = 98$

   $0.7p = 98$

   $2(p - 0.3p) = 98$

   $2(p - 0.3) = 98$

5. **a.** Write an expression for the final price of a property with a base price of p when a 6.5% sales tax *and* an 0.85% property tax are added to the base price.

   b. Let's say that the final price of the property in this situation is $16,639.25. How much is the base price without the taxes?

Percent proportion

Since percentages are fractions, we can easily write proportions to solve percent problems. The basic idea is to write the fraction *part/total* and set that equal to the percentage, which is written as a fraction with a denominator of 100. That way we get this proportion: $\dfrac{part}{total} = \dfrac{percent}{100}$.

Let's solve the problem from the beginning of this lesson using a percent proportion.

Example 2. A handbag has been discounted by 23%, so now it costs $43.81. What was its original price?

The discounted price of $43.18 is the "part" and the original price is the "total" (and is unknown). So we get the fraction $43.81/p. The percentage 77% is written as the fraction 77/100. We get the proportion:

$$\dfrac{\$43.81}{p} = \dfrac{77}{100}$$

Its solution is on the left.

$$\dfrac{\$43.81}{p} = \dfrac{77}{100}$$

$$77p = 100 \cdot \$43.81$$

$$77p = \$4{,}381$$

$$\dfrac{77p}{77} = \dfrac{\$4{,}381}{77}$$

$$p = \$56.90$$

Example 3. Calculate 45% of 0.94 liters.

The total is 0.94 liters, and the part is unknown. We get the percent proportion

$$\dfrac{x}{0.94 \text{ L}} = \dfrac{45}{100}$$

Solve it, and verify that you get $x = 0.423$ liters.

Example 4. The other way to calculate 45% of 0.94 liters is to use decimal multiplication: it is $0.45 \cdot 0.94 \text{ L} = 0.423 \text{ L}$.

(Simply converting percentages into decimals is often more efficient than setting up a percent proportion.)

6. A fan is discounted by 22%, and now it costs $28. Let *p* be its price before the discount.

   a. Find the proportion on the right that matches the problem.

   b. Solve it.

> $28/p = 78/100$
>
> $p/28 = 78/100$
>
> $p/78 = 28$

7. Write and solve a percent proportion (according to the data below) in the form $\dfrac{part}{total} = \dfrac{percent}{100}$.

    a. How much is 56% of 4,500 km?

    b. Thirty-eight percent of a number is 6.08. What is the number?

8. Alice bought 5 bottles of hair conditioner when the store had it at 15% off. Her total bill was $50.79. What was the price of one bottle of hair conditioner before the discount?

9. Matthew has to pay an annual property tax that is 0.8% of the accessed value (the official value for tax purposes) of his land. The tax agency told him the tax is $95.20. From that information, Matthew calculated the accessed value. What is the accessed value?

10. The price of electricity was lowered by 5%, so now it is $0.133 per kilowatt-hour. What was the price before the decrease?

11. A store owner was planning for a big 30% off sale. However, she was rather unethical about it, and the night before the sale, she increased the prices on some of the sale items. For example, she increased the price by 30% for a roll of ribbon that did cost $5. What will the sale price of this roll of ribbon be?

12. The area of a triangular piece of land is 6 square kilometers. If the dimensions, including the base and the altitude, of the triangle were increased by 10%, by how many percent would its *area* increase?

    Hint: Make up a triangle with the given area. In other words, make up a base and an altitude so that the area is 6 km².

Puzzle Corner

A family's water bill for the whole year was $584. From August through December, the bill was 10% higher than from January through July because of a 10% price increase. What was the monthly water bill before the increase?

Circle Graphs

A **circle graph** shows visually how a total is divided into parts (percentages). Each of the parts (pie slices) is a **sector**, and each sector has a **central angle**.

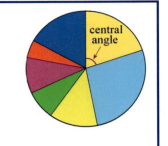

To make a circle graph, we need to calculate the measure of the central angle of each sector. For example, if a circle graph is supposed to show the percentages 25%, 13%, and 62%, simply calculate those percentages of 360° (the full circle):

25% of the total corresponds to 0.25 · 360° = 90°.
13% of the total corresponds to 0.13 · 360° = 46.8°.
62% of the total corresponds to 0.62 · 360° = 223.2°.

1. Sketch a circle graph that shows...

   a. 50%, 25%, and 25%   **b.** 33.3%, 33.3%, 1/6, and 1/6   **c.** 20%, 20%, 10%, and 50%

   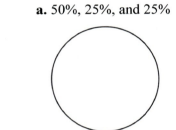 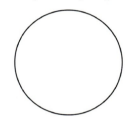 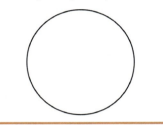

2. The table shows different kinds of specialty breads that a grocery store ordered. Fill in the table. Make a circle graph. (*Note:* You will need a protractor to draw the angles.)

   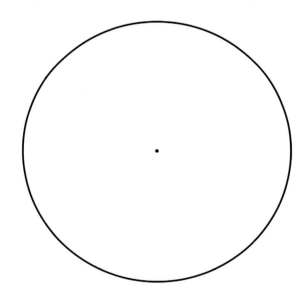

Type	Quantity	Percentage	Central Angle
white bread	50		
bran bread	25		
rye bread	30		
corn bread	40		
4-grain bread	55		
TOTALS	**200**	**100%**	**360°**

3. **a.** Make a bar graph of the quantities of each type of bread from the table above. →

   b. Does the bar graph show percentages?

   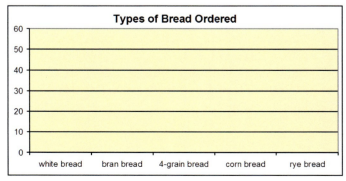

24

4. Think of fractions. Estimate how many percent the sectors of the circle graphs represent.

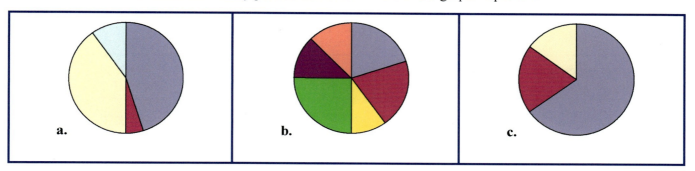

a.

b.

c.

5. The table lists by flavor how many units of protein powder a company sold. Draw a circle graph showing the percentages. You will need a protractor and a calculator.

Flavor	Amount sold	Percentage of total	Central Angle
chocolate	67		
vanilla	34		
strawberry	16		
blueberry	26		
TOTALS		100%	360°

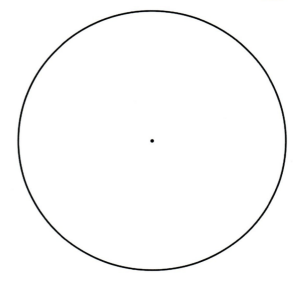

6. Mark polled some seventh graders about their favorite hobbies. Below are his results. Draw a circle graph to show the percentages. Round the angles to whole degrees. You will need a protractor and a calculator.

Favorite hobby	Percentage	Central Angle
Reading	12.3%	
Watching TV	24.5%	
Computer games	21%	
Sports	22.3%	
Pets	7.1%	
Collecting	8.1%	
no hobby	4.7%	
TOTALS	100%	360°

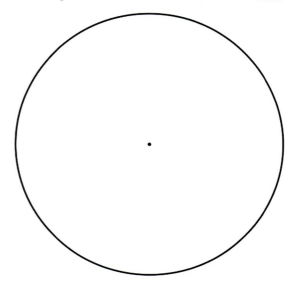

25

Percentage of Change

Percent(age of) change is a way to describe how much a price or some other quantity is increasing or decreasing (changing). Let's look at how to calculate the percentage a quantity is changing.

Example 1. A phone used to cost $50. Now it has been discounted to $45. What percentage was the discount?

Since this problem is asking for the *percentage*, we will use our basic formula $\frac{part}{total} = percentage$.

Because the change is relative to the *original* price, that original price becomes the "total" in our equation. The "part" is the actual amount by which the quantity changes, in this case $5. So we get

$$percentage = \frac{part}{total} = \frac{\$5}{\$50} = 1/10 = 10\%$$

Essentially, we wrote **what fraction the $5 discount is of the original $50 price** and converted that fraction into a percentage.

In summary: To calculate the percent change, use the same basic formula that defines a percentage: *part/total*. Since the change is relative to the original price, the original price is the "total," and the change in price is the "part."

$$percentage\ of\ change = \frac{part}{total} = \frac{difference}{original}$$

1. Write an equation and calculate the percentage of change.

a. A toy construction set costs $12. It is discounted and costs only $8 now. What percentage is the discount? $\frac{difference}{original} =$	**b.** A sewing kit costs $20. It is discounted and costs only $16 now. What percentage is the discount?
c. A bouquet of flowers used to cost $15, but now it costs $20. What is the percentage of increase?	**d.** The price of a stove was $160. The price has increased, and now it costs $200. What is the percentage of increase?

Compare these two problems:	
Gasoline cost $3/gallon last week. Now it has gone up by 5%. What is the new price? 1. Calculate 5% of $3. Since 10% of $3 is $0.30, we know that 5% is half of that, or $0.15. 2. Add $3 + $0.15 = $3.15/gallon. That is the new price.	Gasoline cost $3/gallon last week. Now it costs $3.15. What was the percentage of increase? 1. Find how much was added to $3 to get $3.15 (the difference). That is $0.15. 2. Find what percentage $0.15 is of the original price, $3. It is 15/300 = 5/100 = 5%. So the percentage of increase was 5%.

To find the percentage of increase (the right box above), we work "backwards" compared to when we find the new price when the percentage of increase is known (the left box above).

2. Solve and compare the two problems.

a. A shirt used to cost $24 but it was discounted by 25%. What is the new price?	**b.** A shirt used to cost $24. Now it is discounted to $18. What percentage was it discounted?

3. Solve and compare the two problems.

a. At 5 months of age, a baby weighed 5 kg. At 6 months, the baby weighs 6 kg. What was the percentage of increase?	**b.** At 5 months, a baby weighed 6 kg. Over the next month, his weight increased by 20%. What is his weight at 6 months of age?

4. From June to July, the rent increased from $325 to $342. Then it increased again in August, to $349. Which increase was a greater percentage?

5. **a.** The price of a biology textbook was $60. Then it was lowered to $54.
   Calculate the percentage change in price.

   b. The price was increased back to $60. Calculate the percentage of increase.
   Hopefully this is not a surprise to you, but the percentage is *not* the same as in part (a)!

6. A jacket cost $50. First its price was increased by 20%. Then it was discounted by 20%.

   a. Calculate the final price. It will *not* be $50!

   b. Since the original price was $50, use your answer from part (a) to calculate the true overall percentage change in price.

7. Jake's work hours were cut from 40 to 37.5 a week. Anita's work hours were cut from 145 to 135 a month. Whose work hours were cut by a greater percentage?

8. The price of a vacuum cleaner that cost $100 is increased by 20%. Then it is increased by another 10%.

   a. Find the price of the vacuum cleaner now.

   b. Find the percentage of increase if the price had been increased from $100 to the final price in one single increase. Note: the answer will *not* be 30%!

Percentage of Change: Applications

Area
Example 1. A children's playground measures 30 ft × 40 ft. It is enlarged so that each side is 10 ft longer. What is the percentage of increase in the area? The question doesn't ask for the percent increase of each *side*, but of the *area*. The original area is 30 ft × 40 ft = 1,200 sq. ft. The new area is 40 ft × 50 ft = 2,000 sq. ft. Now we can find the percentage of increase. The fraction *difference/original* is (800 sq. ft. / 1,200 sq. ft) = 8/12. In lowest terms 8/12 becomes 2/3, which, as a percentage, is 66.7%.

Give your answers to the nearest tenth of a percent.

1. Find the percentage of increase in area when a 10 m × 10 m garden is enlarged to be 15 m × 15 m.

2. A newsletter has been printed on 21 cm × 29.7 cm paper. To save costs, it will be printed on 17.6 cm × 25 cm paper instead. By how many percent will the printable area decrease?

3. The sides of a square are increased by a scale factor of 1.15.

   a. By what percentage does the length of each side increase?

   b. What is the percentage of increase in area?
   Hint: Make up a square using an easy number for the length of side.

   c. (Challenge) Would your answers to (a) and (b) change if the shape were a rectangle? A triangle?

4. The graph below compares the production of water in California for usage in June, July, and August of 2013 to the production in the corresponding months of 2014. Because of imposed restrictions on water usage, less water was used in 2014 than in 2013.

   Notice the scale is in millions of gallons. For example, in June, 2013, California produced 215,363 *million* gallons of water, not 215,363 gallons.

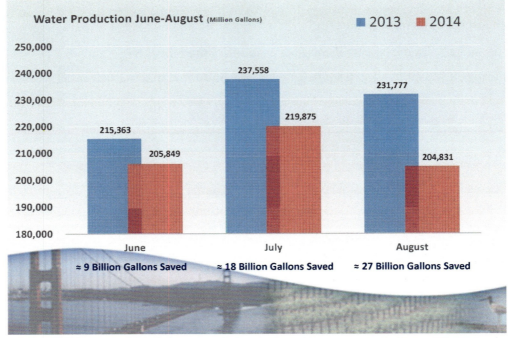

Source: California Water boards

   a. Why do you think the water production (and usage) is higher in July than in June?

   b. Calculate the percentage of decrease between 2013 and 2014 for each of the three months.

   c. In which month was the greatest percentage decrease in water production?

   How can you see that from the graph?

5. The price of a jar of honey went from $5.50 to $6.00. Then it increased further to $6.50. If the price were to increase by another $0.50 (from $6.50 to $7.00), would the percentage increase be more than, less than, or the same as when the price was increased the first time?

6. The sales tax is 7%. The price of a solar battery charger *with tax* is $69.99.

   a. What is the price without tax?

   b. Let's say a merchant wants the price of the battery charger to be increased so that the price including tax is $79.99. What percentage did the price including tax increase?

   c. What is the percentage of increase of the price before tax?

7. A designer plans to use windows of the size 85 cm × 85 cm. He changes his mind and uses windows that are both 10 cm wider and longer instead.

   a. By how many percent does that change the area of one window?

   b. Originally, the designer was going to use 20 of the smaller windows in the house. About how many of the bigger windows cover the same area as 20 of the smaller ones?

8. a. Three items, with prices of $50, $60, and $70, have their prices *increased* by $10. For which item is the *percentage* increase in price the greatest?

   b. Three items, with prices of $50, $60, and $70, have their prices *discounted* by 12%. Which item's price decreases the most (in dollars)?

9. The population of the state of Kentucky was 3,038,000 in 1960 and 3,219,000 in 1970. Calculate the percentage of increase in the population of Kentucky during that decade.

10. The table shows the population of Kentucky every 10 years. The line graph shows the same information. Your task is to calculate the percentage of increase in each decade. You already did that for 1960-1970 in the previous exercise, so write that answer in the row for 1970.

Year	Population	% increase in the decade
1960	3,038,000	N/A
1970	3,219,000	
1980	3,661,000	
1990	3,685,000	
2000	4,042,000	
2010	4,340,000	

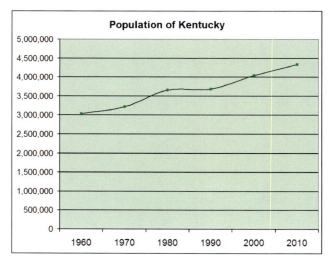

11. The population of Kentucky grew more steeply in one of these decades than any of the others.

   a. Which decade? From _____ to _____.

   b. How can you tell which decade it is from the graph?

12. (Optional.) Make a line graph that shows the change in the population where you live (your town, city, parish, county, state, province, country, etc.). Also calculate the percentage of increase or decrease for each time period. You can often find population statistics in Wikipedia, for example. You can also search the Internet (with adult supervision) for "your area population statistics," where your area is the place that you chose.

Comparing Values Using Percentages

What percentage more/less/bigger/smaller/taller/shorter ...?

Example 1. A car weighs 2,000 kg. Another car weighs 2,500 kg. The second car is heavier, but how much heavier than the first one is it?

To answer that question, we cannot just look at the difference of 500 kg and say that 500 kg is "a lot" or "a little." Instead, we need to take into account the sizes of the two items being compared and consider their *relative difference*.

The **relative difference** (or **percentage of difference**) of two values is the fraction $\frac{\text{actual difference}}{\text{reference value}}$.

This fraction is usually expressed as a percentage.

The problem is in determining which of the two values is the reference value. In this case, we want to find out how much heavier the second car is than the first car, which means we use the weight of the first car as the reference value. In another words, we are comparing the second car to the first car. It is as if the lighter car were there first, and we are comparing a "newcomer" car to this first car.

So the relative difference of the two weights is $\frac{500 \text{ kg}}{2{,}000 \text{ kg}} = \frac{5}{20} = \frac{1}{4} = 25\%$.

This means the second car is 25% heavier than the first one. This gives us a precise value.

Choosing the second car's weight as the reference value gives: $\frac{500 \text{ kg}}{2{,}500 \text{ kg}} = \frac{5}{25} = \frac{1}{5} = 20\%$.

This means the first car is 20% lighter than the second car.

Example 2. Southcreek College has 2,600 students and West River College has 2,400 students. How many percent more students does Southcreek College have than West River College?

The difference in student count is 200. But which number is our reference? Since West River College is mentioned after the word "than," we are comparing Southcreek College to West River College. So the student count of West River College is our reference number.

The fraction is $\frac{\text{difference in student count}}{\text{reference student count}} = \frac{200}{2{,}400} = \frac{2}{24} = \frac{1}{12} = 0.08333... \approx 8.3\%$.

So Southcreek College has approximately 8.3% more students than West River college.

1. Compare the taller object to the shorter one and calculate their relative difference.

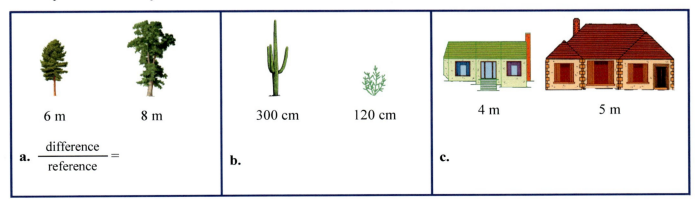

a. $\frac{\text{difference}}{\text{reference}} =$

b.

c.

You may use the calculator in all problems in this lesson from this point on.

2. Erica is 140 cm tall, and Heather is 160 cm tall. Fill in the blanks.

   Heather is ─── = _____% taller than Erica. Erica is ─── = _____% shorter than Heather.

3. **a.** Refer to the chart. Calculate the percentage differences to the tenth of a percent.

   The population of Tokyo is _____% larger than the population of Delhi.

   The population of Moscow is _____% smaller than the population of Mexico City.

   b. Compare how much larger the population of Seoul is than the population of New York, and then again how much larger the population of New York is than the population of Moscow.

   Which is a larger percentage difference?

Metropolitan area	Country	Population (millions)
Tokyo	Japan	37.80
Seoul	South Korea	25.62
Shanghai	China	24.75
Karachi	Pakistan	23.50
Delhi	India	21.75
Mexico City	Mexico	21.60
Sao Paulo	Brazil	21.20
Jakarta	Indonesia	20.00
New York City	United States	19.95
Mumbai	India	20.75
Moscow	Russia	15.51

> When there is no clear way to choose a reference value when calculating the relative (percentage) difference, you can use the average of the two values as a reference value.
>
> **Example 3.** The monthly subscription fees to two math practice websites are $9.99 and $12.95. What is their relative difference?
>
> The average of the two prices is ($9.99 + $12.95)/2 = $11.47.
>
> The relative difference is therefore $\dfrac{\$12.95 - \$9.99}{\$11.47} = \dfrac{2.96}{11.47} \approx 25.8\%$.

4. The area of one park is 14,000 square feet, and the area of another park is 10,000 square feet. Use their average area to calculate the relative percentage difference between their areas.

5. KeepCool company charges $28 per hour for labor, and CityFreez charges $32 per hour.

   a. Use the average rate for labor to calculate the percentage difference in the rates.

   b. Now compare the costs for *two* hours of labor. What is the percentage difference?

Example 4. One cat weighs 1.2 kg and another weighs 1.5 kg.

Question 1. What percentage is the smaller cat's weight <u>of</u> the bigger cat's weight?

Solution: Write what fraction the smaller cat's weight is relative to the bigger cat's weight:

It is $\dfrac{\text{smaller cat's weight}}{\text{bigger cat's weight}} = \dfrac{1.2 \text{ kg}}{1.5 \text{ kg}} = \dfrac{12}{15} = \dfrac{4}{5} = 80\%.$

> Compare carefully these five types of questions about the same situation!

Question 2. What percentage is the bigger cat's weight <u>of</u> the smaller cat's weight?

Solution: Write what fraction the bigger cat's weight is relative to the smaller cat's weight:

It is $\dfrac{\text{bigger cat's weight}}{\text{smaller cat's weight}} = \dfrac{1.5 \text{ kg}}{1.2 \text{ kg}} = \dfrac{15}{12} = \dfrac{5}{4} = 1\,¼ = 125\%.$

Question 3. How much more (in percent) does the bigger cat weigh <u>than</u> the smaller cat?

Solution: This is a percentage difference relative to the smaller cat. Write the fraction (difference / reference):

It is $\dfrac{\text{difference in weight}}{\text{smaller cat's weight}} = \dfrac{0.3 \text{ kg}}{1.2 \text{ kg}} = \dfrac{3}{12} = \dfrac{1}{4} = 25\%.$

Question 4. Compare the weight of the smaller cat <u>to</u> the weight of the bigger cat.

Solution: This percentage difference is relative to the weight of the bigger cat:

It is $\dfrac{\text{difference in weight}}{\text{bigger cat's weight}} = \dfrac{0.3 \text{ kg}}{1.5 \text{ kg}} = \dfrac{3}{15} = \dfrac{1}{5} = 20\%.$

Question 5. What is the relative difference between the two cats' weights?

Solution: No reference is specified, so let's use the average weight: ½(1.5 kg + 1.2 kg) = 1.35 kg.

It is $\dfrac{\text{difference in weight}}{\text{average weight}} = \dfrac{0.3 \text{ kg}}{1.35 \text{ kg}} = \dfrac{30}{135} = \dfrac{2}{9} = 0.222\ldots \approx 22.2\%.$

There are five different questions with five different solutions. Note the underlined key words!

You could be asked, **"What percentage is (this) <u>of</u> (that)?"**

OR: **"What percentage <u>more/less/smaller/bigger</u> is (this) <u>than</u> (that)?"**

Moreover, the order of comparison matters: The keywords "of," "than," and "to" mark the cat that we are comparing to (the reference cat).

---

6. One bean plant is 12 cm tall, and another is 16 cm tall.

a. What percentage is the shorter plant's height of the taller plant's height?	**b.** How much taller (in percent) is the taller plant than the shorter plant?
c. Compare the height of the shorter plant to the height of the taller one.	**d.** Find the relative difference in the heights using their average height.

7. Only one of the students gets the answer correct in each case. Who is it?

   a. Jack made a tower of blocks 150 cm tall. Baby made a block tower that was 30 cm tall. What percentage is the height of Baby's tower of the height of Jack's tower?

Elijah:	**Angela:** I write the fraction	**Mary:** I write the fraction
I subtract 150 − 30 = 120%.	$\dfrac{120 \text{ cm}}{150 \text{ cm}} = \dfrac{12}{15} = \dfrac{4}{5} = 80\%$	$\dfrac{30 \text{ cm}}{150 \text{ cm}} = \dfrac{3}{15} = \dfrac{1}{5} = 20\%$

   b. The school orchestra has 26 boys and 14 girls. How many percent bigger is the number of boys than the number of girls?

Elijah: I subtract 26 − 14 = 12 and write the fraction $\dfrac{12}{14} = \dfrac{6}{7} \approx 86\%$	**Angela:** I subtract 26 − 14 = 12%.	**Mary:** I write the fraction $\dfrac{14}{26} = \dfrac{7}{13} \approx 54\%$

8. Percentage comparisons can be misleading without the actual data. To see that, consider the number of violent crimes committed in two imaginary counties in 2013 and 2014.

   a. Calculate the percentage of increase in violent crime for the two counties in the table.

   b. Let's say you were thinking of moving into one of these two counties, and you were told *only* the percentages of change for violent crime. Which county might you move into?

   Number of violent crimes in 2013-2014

	County A	County Z
2013	2	454
2014	4	512
Percentage of increase		

Yet, taking the *actual data* into account, your perception about these two areas would change a lot!

The table lists the number of domestic burglaries in three counties in Wales for two different time periods.

Number of domestic burglaries

	Ceredigion	Conwy	Gwynedd
April 2011 to March 2012	93	256	200
April 2012 to March 2013	56	229	138
Approximate percent of change			

Determine *without* a calculator or paper and pencil (just figure in your head) which county experienced the greatest percentage of decrease in domestic burglaries.

Simple Interest

When you deposit money into a savings account, normally the bank actually pays you for the use of your money. The amount you deposit is called the **principal**. The amount they pay is called the **interest**.

Similarly, if you borrow money, it is not free. You not only pay back the principal (the actual amount you borrowed) but you also pay interest on the amount of money you borrowed.

Interest is normally defined as a certain percentage, called the **interest rate**, of the principal. For example, the bank might charge you an interest rate of 7.9% on a loan. Often, people simply say "7.9% interest" instead of "a 7.9% interest rate" or "an interest rate of 7.9%."

In this lesson, we study only so-called **simple interest**, which means that the interest is added to the principal only at the end of the time period during which the money is invested or borrowed. In real life, banks usually calculate **compound interest**, which means that the interest is added to the principal at certain regular intervals (such as every month or even every day) during the period of the loan or investment.

Example 1. How much interest does a principal of $2,000 earn in a year if the yearly interest rate is 5%?

The interest is simply 5% of $2,000, which is $100.

Example 2. You get a $3,000 loan with an annual interest rate of 8.5%.
You pay the loan back after 3 years. How much do you have to pay back?

The interest for one year is simply 8.5% of $3,000. For three years, it is three times that much.
Of course you also have to pay back the of principal $3,000. So the total amount you pay back is:

$3,000 + 3 · 0.085 · $3,000 = $3,000 + $765 = $3,765.

We can see from the above examples that to calculate the interest, we simply take the interest rate times the principal times the time period. The formula for calculating simple interest (*I*) is usually given as:

$$I = Prt$$

where *P* is the principal, *r* is the interest rate, and *t* is the time.

You may use a calculator in all problems in this lesson.

1. Calculate the interest and the total amount to be paid back on these investments.

    a. Principal $5,000; interest 3%; time 1 year

    Interest: _____ Total to withdraw: _____

    b. Principal $3,500; interest 4.3%; time 4 years

    Interest: _____ Total to withdraw: _____

    c. Principal $20,000; interest 7.6%; time 10 years

    Interest: _____ Total to withdraw: _____

2. Sandy plans to invest $3,000 for three years. She could either put it into a savings account with an annual interest rate of 3.4% or get a Certificate of Deposit (CD) for 5 years with an annual interest rate of 3.92%. However, if she withdraws the money from the CD before 5 years is up, the bank charges her a penalty of 6 months interest. Which allows Sandy to earn more money on her investment in 3 years?

Savings Account

Interest rate: 3.4%

Certificate of Deposit

Time period: 5 years

Interest rate: 3.92%

Example 3. Andrew borrows $2,000 with an annual interest rate of 12.45%. He is able to pay it back after 7 months. How much will Andy pay to the lender?

Notice that the interest rate is *annual* (yearly) but the time period is in *months*. Therefore:

(1) We need to either use a monthly interest rate. To get that, simply divide the annual interest rate by 12.

(2) Or we need to convert the time of 7 months into years. Of course, 7 months is simply 7/12 years.

Let's use the first option. The monthly interest rate is 12.45% ÷ 12 = 1.0375%. Then, the interest is $0.010375 \cdot 7 \cdot \$2{,}000 = \$145.25$. So Andy has to pay back $2,145.25.

---

3. Elizabeth bought a tablet for $450 on credit with a 12.9% annual interest rate.

   a. How much interest (in dollars) will she pay in a month?

   b. In a day?

4. A credit card has a monthly interest rate of 1.09%, which doesn't sound like much. How much interest will you pay if you purchase a couch for $690 with the credit card and pay it back after two years?

5. Jerry took out a loan for $850 for 10 months with an annual interest rate of 10.8%. How much less interest would he have paid if instead he had taken out a loan for 7 months with an annual interest rate of 9.5%?

6. John uses his credit card to finance a car for $26,000. The annual interest rate on his card is low, 2.75%, but only for the first 12 months. After that, if Jon hasn't paid the full amount back, the annual interest rate jumps to 9.95%. Calculate how much John ends up paying back if he cannot pay the total during the first 12 months, but pays the entire amount after 2.5 years.

Example 4. Eric borrowed $1,500 for 8 months and paid back $1,578. What was the interest rate?

The actual interest was $1,578 − $1,500 = $78. The loan was for 8 months, so the monthly interest was $78 ÷ 8 = $9.75.

This amount is $9.75/$1,500 = 0.0065 = 0.65% of the principal. So the monthly interest rate was 0.65%. The annual interest rate is 12 times that, or 7.8%.

Another solution. We could also write an equation, using the formula $I = Prt$.

This formula is for the actual amount of interest (I), not the total Eric paid back. The actual interest (I) was $78. We also know that the time was 8 months and that the principal was $1,500. Substituting those values into the formula and keeping the units, we get

$$\$78 = \$1,500 \cdot r \cdot 8 \text{ months}$$

It helps to write the equation so that the variable is on the left (inverting the sides):

$$\$1,500 \cdot r \cdot 8 \text{ months} = \$78$$

To solve this equation for r, we divide both sides by $1,500 and then also by 8 months. See the solution on the right.

$$\$1,500 \cdot r \cdot 8 \text{ months} = \$78$$

$$\frac{\$1,500 \cdot r \cdot 8 \text{ months}}{\$1500} = \frac{\$78}{\$1500}$$

$$r \cdot 8 \text{ months} = \frac{78}{1500}$$

$$\frac{r \cdot 8 \text{ months}}{8 \text{ months}} = \frac{\frac{78}{1500}}{8 \text{ months}}$$

$$r = \frac{0.0065}{\text{month}}$$

$$r = 0.65\% \text{ per month}$$

7. **a.** You borrowed $1,000 for one year. At the end of the year, you had to pay back $1,045. What was the interest rate?

   b. You borrowed $12,000 for five years. At the end of the five years you paid the bank back $15,600. What was the interest rate?

8. Alice opened a savings account that paid an interest rate of 6%. After ten years her account contained $12,000. How much was the original principal?

9. How long would you have to invest $2,000 in order to earn $500 in interest, if the annual interest rate is 11.5%?

10. What rate of interest do you need in order to earn $350 in interest in 2 years with a principal of $1,800?

11. A family purchased a vacation package for $4,055 with a credit card that charged 11.95% annual interest. When they paid off their credit card, they ended up paying the credit card company $4,741.48. How long did it take for the family to pay off the debt for their vacation?

Puzzle Corner

Let's look at just one example of how **compound interest** is calculated! Jayden buys a motorcycle for $5,000 with a credit card that has a 6% annual interest rate. Compounding the interest means that the interest is added to the principal at certain intervals, in this case each month.

Study the table below. Note that a 6% annual interest rate means 0.06/12 = 0.005 or 0.5% monthly interest rate.

a. Fill in the table, calculating the interest for each month and adding it to the principal. The following month, use the new principal to calculate the interest.

Month	Monthly interest	Principal
0		$5,000
1	$5,000 · 0.005 = $25	$5,025
2	$5,025 · 0.005 = $25.125	$5,050.125
3	$5,050.125 · 0.005 ≈ $25.251	$5,075.376
4	_____ · 0.005 ≈	
5	_____ · 0.005 ≈	
6		
7		
8		

b. There is a formula for compound interest that allows these calculations to be done quicker. For this situation, the total amount owed after month n is

$$\$5{,}000 \cdot 1.005^n$$

For example, after 12 months, Jayden owes $5,000 · 1.005^{12} = $5,308.39. Use the formula to calculate how much Jayden pays if he pays back the loan at the end of 2 years.

Review

1. Find the percentage of the given quantity.

   a. 9.2% of $150   **b.** 45.8% of 16 m   **c.** 0.6% of 700 mi

2. All these items are on sale. Calculate the new, discounted prices.

   a. Price: $9
   20% off
   New price: $_____

   b. Price: $6
   25% off
   New price: $_____

   c. Price: $90
   30% off
   New price: $_____

3. A flashlight is discounted by 18%, and now it costs $23.37. Let p be its price before the discount. Find the equation that matches the statement above and solve it.

   $p - 0.18 = \$23.37$

   $p - 18 = \$23.37$

   $0.82p = \$23.37$

   $0.18p = \$23.37$

4. Two brothers, Andy and Jack, are sharing the price of a new computer so that Andy pays 2/5 (or 40%) and Jack pays 3/5 (or 60%) of the price. The computer costs $459, and there is a sales tax of 7% that will be added to the price. Calculate Andy's and Jack's shares.

5. A portable reading device costs $180. Now it is discounted and costs $153. What was the percentage of discount?

6. The two right triangles on the right are similar.

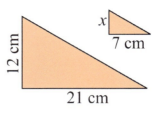

   a. Calculate what percentage the area of the smaller triangle is of the area of the larger triangle.

   b. In what ratio are the corresponding sides of the triangles?

   c. In what ratio are their areas?

7. A wall painting was planned to be 5 m by 3 m in size. If both of its sides are increased by 20%, by what percentage will the area of the painting increase?

8. In a race, Old Paint finished in 120 seconds, and The Old Gray Mare finished in 200 seconds.

   a. How many percent quicker was Old Paint than The Old Gray Mare?

   b. How many percent slower was The Old Gray Mare than Old Paint?

   c. What is the relative difference in their times?

9. Noah takes a $4,000 loan at a 7.8% annual interest rate to purchase a car. At the end of the first year, he pays back $2,000 of the principal of the loan. At the end of the second year, he pays back the rest of the principal and all of the interest. How much total interest does he have to pay? Assume simple interest, which is calculated and paid only at the end of the period of the loan.

Percent, Grade 7 Answer Key

Review: Percent, p. 10

1.

a. 52% = $\frac{52}{100}$ = 0.52	b. 7% = $\frac{7}{100}$ = 0.07	c. 59% = $\frac{59}{100}$ = 0.59
d. 109% = $\frac{109}{100}$ = 1.09	e. 382% = $\frac{382}{100}$ = 3.82	f. 200% = $\frac{200}{100}$ = 2.0

2.

a. 28.2% = $\frac{282}{1000}$ = 0.282	b. 6.7% = $\frac{67}{1000}$ = 0.067	c. 89.1% = $\frac{891}{1000}$ = 0.891
d. 0.9% = $\frac{9}{1000}$ = 0.009	e. 10.39% = $\frac{1039}{10000}$ = 0.1039	f. 340.9% = $\frac{3409}{1000}$ = 3.409
g. 45.39% = 0.4539	h. 2.391% = 0.02391	i. 94.2834% = 0.942834

3.

a. $\frac{8}{25}$ = $\frac{32}{100}$ = 32%	b. $\frac{142}{200}$ = $\frac{71}{100}$ = 71%	c. $\frac{24}{20}$ = $\frac{120}{100}$ = 120%

4.

a. 11/8 = 137.5%	b. 11/24 ≈ 45.8%
| ```
     1.3 7 5
  8)1 1.0 0 0
    - 8
      3 0
     -2 4
        6 0
       -5 6
          4 0
         -4 0
            0
``` | ```
      0.4 5 8 3
  24)1 1.0 0 0 0
    - 9 6
      1 4 0
     -1 2 0
        2 0 0
       - 1 9 2
            8 0
           - 7 2
              8
``` |

5.

a. $\frac{2}{3}$ ≈ 0.6667 = 66.67%	b. $\frac{6}{7}$ ≈ 0.8571 = 85.71%
c. $\frac{17}{23}$ ≈ 0.7391 = 73.91%	d. $\frac{52}{98}$ ≈ 0.5306 = 53.06%

6.

Ocean	Area / Total Area	Percent
Pacific	155,557,000 km^2 / 335,258,000 km^2 = 0.46399...	46.4%
Atlantic	76,762,000 km^2 / 335,258,000 km^2 = 0.22896...	22.9%
Indian	68,556,000 km^2 / 335,258,000 km^2 = 0.20448...	20.4%
Southern	20,327,000 km^2 / 335,258,000 km^2 = 0.06063...	6.1%
Arctic	14,056,000 km^2 / 335,258,000 km^2 = 0.04192...	4.2%

Review: Percent, cont.

7. The area of the Carters' land is 85.5% of that of the Joneses'.
   The area of the Carters' land is 40 m · 35 m = 1,400 m². The area of the Joneses' land is 42 m · 39 m = 1,638 m².
   So the area of the Carters' land is 1,400/1,638 ≈ 0.8547 ≈ 85.5% of the area of the Jones's land.

8. a. The weight of Captain is 140% of the weight of Chief.
   b. Captain weighs 140% · 6 lb = 1.4 · 6 lb = 8.4 lb.

Puzzle corner.

a. There are two quarters fully uncolored and there is one quarter fully colored. The fourth quarter is half colored. (Count the little squares: $^8/_{16}$ = ½). So one full quarter (two-eighths) and half of another quarter (one more eighth) are colored. As a percentage, 3/8 = 0.375 = 37.5%.

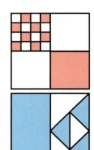

b. The blue rectangle on the left is half of the square. The other rectangle on the right is also half of the square. In that rectangle on the right, the four small triangles have the same area as the two larger ones. (Imagine that the entire square is divided into quarters.) So the three blue triangles make up ¾ of half of the area of the right rectangle, so they make up ¾ of ½ of ½ of the entire square. As a percent, the blue-shaded area is the sum of the triangles and the rectangle:
(¾ · ½ · ½) + (½) = $^3/_{16}$ + $^8/_{16}$ = $^{11}/_{16}$ = 0.6875 = 68.75%.

Solving Basic Percent Problems, p. 13

1.

| a. 107% = 1.07 | b. 16.67% = 0.1667 | c. 4.5% = 0.045 |

2. a. Laptop, original price $249.90, increase 6%.
   New price = 1.06 · $249.90 = $264.90.

   b. Biology textbook, original price $82.40, increase 2.5%.
   New price = 1.025 · $82.40 = $84.46.

   c. Lunch buffet, original price $18.50, increase 11.2%.
   New price = 1.112 · $18.50 = $20.57.

3. Part / Total = $325.08 / $1,890 = 0.172 = 17.2%. Julia paid 17.2%.

4. The new price is $39.59.

   > Since 13% of the price is removed, 87% of the price is left. I write that percentage as a decimal and multiply the original price by it: 0.87 · $45.50 = $39.59.

5. a. New price = 0.62 · $78 = $48.36
   b. New price = 0.775 · $14.95 = $11.59

6. a. 9 in : 18 in = 1:2.

   b. Since the sides are in a ratio of 1:2, the unknown side of the larger rectangle is 30 in.
      So the area of the smaller rectangle is 15 in · 9 in = 135 sq in and of the larger is 30 in · 18 in = 540 sq in.
      Thus the two areas are in a ratio 135 : 540 = 1:4. (Notice that the ratio of the areas is just the square of the ratio of the sides: ½ · ½ = ¼.)

   c. The area of the smaller rectangle is 135 cm² / 540 cm² = 25% of the area of the larger rectangle.

7.

| a. Bicycle: $100; 7% sales tax.
Tax to add: $7
Price after tax: $107 | b. Fridge: $400; 6% sales tax.
Tax to add: $24
Price after tax: $424 | c. Haircut: $50; 3% sales tax.
Tax to add: $1.50
Price after tax: $51.50 |

Solving Basic Percent Problems, cont.

8.

a. A microscope.	b. 2 shirts.	c. Crayons and a toy horse.
$180 + $9 = $189	2($18 + $0.90) = $37.80	$2 + $13 = $15 $15 + $0.75 = $15.75
d. A bag and a shirt.	e. 6 bottles of water.	f. 2 shirts and a bag.
$70 + $18 = $88 $88 + $4.40 = $92.40	6 · $2 = $12 $12 + $0.60 = $12.60	$18 + $18 + $70 = $106 $106 + $5.30 = $111.30

9. The final price of a music CD is $11.50 + $0.77 = **$12.27**.

10. Jeremy will pay $850 · (1 − 0.375) = **$531.25** for the vacation package.

11. a. The total price Patrick paid was 1.012 · $200,000 = $202,400.
    b. 2,000 m² / 5,000 m² = 2/5 = 40%; 0.4 · $202,400 = $80,960.
       Patrick should charge his neighbor $86,400 in order to recover what he paid for the parcel that he's selling him.

12. a. The final price of the CD is 0.8 · $12.50 + 0.07 · $10 = **$10.70**.
    b. The final price of the jeans is 0.6 · $55.97 + 0.05 · $33.58 = **$35.26**.

13. a. The average price of the four types of pasta is **$1.85**.
    b. The average price would be **$1.67** after the discount.

Percent Equations, p. 16

1. a. headphones, price $12, discounted by 24%. New price = <u>0.76 · $12</u>
   b. a bag of dog food, price p, discounted by 11%. New price = <u>0.89p</u>
   c. pizza sauce, price x, discounted by 17%. New price = <u>0.83x</u>
   d. sunglasses, price s, price increased by 6%. New price = <u>1.06s</u>

2. $p - 0.25p$ = $576
    0.75p = $576 | ÷ 0.75
    p = $768

    The computer cost **$768** before the discount.

3. $r + 0.05r$ = $215.25
    1.05r = $215.25 | ÷ 1.05
    1.05r / 1.05 = $215.25 / 1.05
    r = $205.00

    The rent was **$205** before they raised it.

4. a. 2(p − 0.3p) = 98

   b. 2(p − 0.3p) = $98 | ÷ 2
       2(p − 0.3p) / 2 = $98/2
       p − 0.3p = $49
       0.7p = $49 | ÷ 0.7
       p = $49/0.7 = $70

   The original price was **$70 per tablet**.

Percent Equations, cont.

5. a. $(1 + 0.065 + 0.0085)p$ which simplifies to $\underline{1.0735p}$.

   b. $\quad 1.0735p\ =\ \$16{,}639.25 \quad\ |\ \div 1.0735$

   $\qquad\qquad p\ =\ \$16{,}639.25 / 1.0735$

   $\qquad\qquad p\ =\ \$15{,}500$

6. a. The percent proportion is *part/total* = *percent*/100. In this case, we have (discounted price)/(original price) = 78/100, so the correct equation is **28/p = 78/100.**

   b. $\dfrac{\$28}{p} = \dfrac{78}{100}$

   $78p = 100 \cdot \$28$

   $78p = \$2{,}800$

   $\dfrac{78p}{78} = \dfrac{\$2{,}800}{78}$

   $p = \underline{\$35.90}$

7. a. $\dfrac{d}{4{,}500\text{ km}} = \dfrac{56}{100}$

   $100d = 56 \cdot 4{,}500 \text{ km}$

   $100d = 252{,}000 \text{ km}$

   $\dfrac{100d}{100} = \dfrac{252{,}000 \text{ km}}{100}$

   $d = \underline{2{,}520 \text{ km}}$

   b. Because here the unknown is the whole (100%), the value 6.08 is the part (38%). Thus the 6.08 and the 38% both go on the top of the proportion, and the unknown number n and the 100% go on the bottom, like this:

   $\dfrac{6.08}{n} = \dfrac{38}{100}$

   $38n = 100 \cdot 6.08$

   $38n = 608$

   $\dfrac{38n}{38} = \dfrac{608}{38}$

   $n = \underline{16}$

Percent Equations, cont.

8. If the conditioner was 15% off, it cost 100% − 15% = 85% of its regular price. Also, with the discount a single bottle cost $50.79 / 5 = $10.158. So the proportion is (discounted price)/(regular price) = 85/100. Let b be the regular price of one bottle. Then the proportion becomes:

$$\frac{\$10.158}{b} = \frac{85}{100}$$

$$85b = 100 \cdot \$10.158$$

$$85b = \$1015.80$$

$$\frac{85b}{85} = \frac{\$1015.80}{85}$$

$$n \approx \$11.95$$

The price of one bottle of hair conditioner before the discount was $11.95. You can of course solve this problem in other ways, as well.

9. The part is the tax, and the total is the value, v. The percent proportion is $95.20/v = 0.8/100$. You can also solve this by writing the equation $0.008v = \$95.20$. Either way, the solution is $v = \$11,900$.
The tax agency valued his land at $11,900.

10. If the price of electricity was lowered by 5%, it is now 95% of what it was before, so we can write a percent proportion now/then = $0.133 / p = 95/100$. Or, you can write the equation $0.95p = \$0.133$. Either way, the solution is $p = \$0.140$.
The price of electricity was $0.14 per kilowatt-hour before the decrease in cost.

11. The price after the discount ends up being $4.55.
The price after the 30% increase is 130% of the original price: $5 · 1.3 = $6.50. The price after the 30% discount is 70% of that higher price: $6.50 · 0.7 = $4.55. The prices are different because the 30% discount affects a bigger price than the 30% increase did (it is a discount not only on the original price but also on the increase), so the discount ends up being bigger than the increase. In this case, the customer got only a $0.45/$5.00 = 9% discount from the original $5.00 price.

12. Answers will vary. Please check the student's work.

    Substituting numbers: Suppose the base b were 3 km and the height h were 4 km to give A = ½ bh = ½ (3 km · 4 km) = 6 km². Increasing 3 km and 4 km by 10% gives the increased area A' = ½ (3.3 km · 4.4 km) = ½ (14.52 km²) = 7.26 km², which is an increase of 1.26 km / 6 km = 21%.

    Using algebra: The formula for the area of a triangle is A = ½ bh = 6 km², where A is the 6-km² area, b is the base, and h is the height. In this case we have A' = ½ b'h', where A' is the area of the increased plot, b' its base, and h' its height. Because the increase is 10%, we know that $b' = 1.1b$ and $h' = 1.1h$. If we substitute into the formula for A', we get A' = ½ b'h' = ½ (1.1b)(1.1h) = 1.21 (½ bh). Since A = ½ bh = 6 km², we get A' = 1.21 A = 7.26 km². The size of the increased plot is 121% of the original plot, so the increase was 21%.

    Using logic: The easiest way to solve this problem is with a little insight: We see that when we increase b and h by 10% in A = ½ bh, the new b and h are 110% or 1.1 times the old ones. Since the ½ in the formula doesn't change, the new area is just 1.1 · 1.1 = 1.21 times the old area, so the increase is 21%.

Puzzle Corner: Let's use m to represent the monthly water bill before the increase. The bill after the 10% increase is $1.1m$. For 7 months, January through July, the bill is m, for a total of $7m$. For 5 months, August through December, the bill is $1.1m$, for a total of $5 \cdot 1.1m = 5.5m$. The total of the bills for the whole year is thus $7m + 5.5m = 12.5m = \$584.00$. Therefore the monthly water bill before the increase was: $m = \$584.00 / 12.5 = \$46.72$.

Circle Graphs, p. 21

1. a. 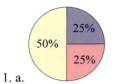 b. 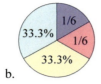 c.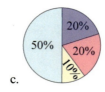

2.

Type	Quantity	Fraction	Percent	Central Angle
white bread	50	1/4	25%	90°
bran bread	25	1/8	12.5%	45°
rye bread	30	3/20	15%	54°
corn bread	40	1/5	20%	72°
4-grain bread	55	11/40	27.5%	99°
TOTAL	200	1	100%	360°

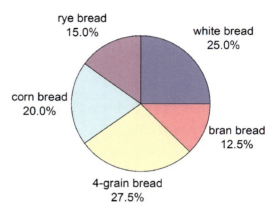

3. a.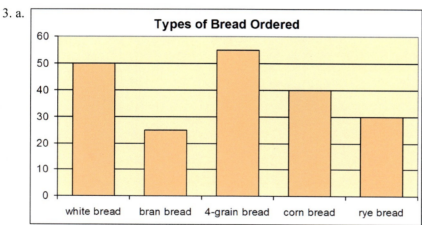

   b. No, it does not.

4. Answers will vary. The answers below are actually the *exact* percentages. However, students' answers may differ from these a little and be totally acceptable. (Each circle should sum to 100%.) Starting from 12:00 and going clockwise:
   a. About 45%, about 5%; about 40%, about 10%. (The first two and the last two should each sum to 50%.)
   b. About 20%, 20%, 10%; 25%, 12.5%, 12.5%. (The first three and the last three should each sum to 50%.)
   c. About 65%, about 20%, about 15%.

5.

Flavor	Amount sold	Percent of total	Central Angle
chocolate	67	46.9%	169°
vanilla	34	23.8%	86°
strawberry	16	11.2%	40°
blueberry	26	18.2%	65°
TOTAL	143	100%	360°

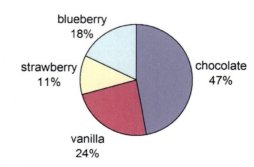

Note: The "Percent of total" column actually totals to 100.1% because the numbers that were rounded up moved a little bit farther in the upward direction than the numbers that were rounded down moved in the downward direction. That's typical for calculations that round several numbers.

Circle Graphs, cont.

6.

Favorite hobby	Percent	Central Angle
Reading	12.3%	44°
TV	24.5%	88°
Computer games	21%	76°
Sports	22.3%	80°
Pets	7.1%	26°
Collecting	8.1%	29°
no hobby	4.7%	17°
TOTAL	100%	360°

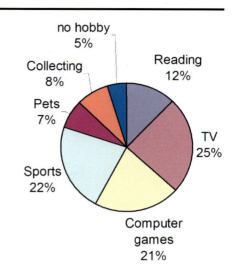

Percentage of Change, p. 23

1. a. Discount: *difference/original* = ($12 − $8) / $12 = $4 / $12 = $0.\overline{3}$ = 33.3%.
   b. Discount: *difference/original* = ($20 − $16) / $20 = $4 / $20 = 0.2 = 20%.
   c. Increase: *difference/original* = ($20 − $15) / $15 = $5 / $15 = $0.\overline{3}$ = 33.3%.
   d. Increase: *difference/original* = ($160 − $200) / $160 = $40 / $160 = 0.25 = 25%.

2. a. Since 25% of $24 is $6, the discounted price is $24 − $6 = $18.
   b. The difference is $24 − $18 = $6, so the percentage of discount is $6/$24 = 0.25 = 25%.

3. a. The difference is 6 kg − 5 kg = 1 kg, so the percentage of increase is 1 kg/5 kg = 0.20 = 20%.
   b. Since 20% of 6 kg is 1.2 kg, his weight at six months is 6 kg + 1.2 kg = 7.2 kg.

4. The difference from June to July is $342 − $325 = $17, and the percentage change is measured relative to the rent for the month of June. So the percentage of increase is $17/$325 ≈ 0.052 = 5.2%. The difference from July to August is $349 − $342 = $7, and the percentage change is measured relative to the rent for July. So the percentage of increase is $7/$342 ≈ 0.020 = 2.0%. Therefore, the change from June to July is a greater percentage increase than the change from July to August.

5. a. The change is $60 − $54 = $6, so the percentage decrease is $6/$60 = 0.10 = 10%.
   b. The change is still $60 − $54 = $6, but this time the percentage of increase is measured from the lower price. So the percentage of increase is $6/$54 = $0.\overline{1}$ ≈ 11.1%. The reason the percentage changes are not the same is that we are measuring them relative to different prices.

6. a. Calculate the final price: Since 20% of $50 is $10, after the increase the price was $50 + $10 = $60. The following decrease in price is measured relative to that $60. Since 20% of $60 is $12, the final price is $60 − $12 = $48.
   b. Calculate the true percentage change: Since $50 − $48 = $2, the true percentage change is a decrease of $2/$50 = 0.04 = 4%. The final price is lower because the 20% discount ended up being larger than the 20% increase. It was larger because it was calculated relative to a higher price.

7. Jake's hours were cut by (40 − 37.5)/40 = 0.0625 = 6.25%. Anita's were cut by (145 − 135)/145 ≈ 0.069 = 6.9%. So Anita's hours were cut by a larger percentage.

8. a. The first increase (by 20%) raised the price to 1.2 · $100 = $120. The second increase (by 10%) started from $120, so it raised the price to 1.1 · $120 = <u>$132</u>.

   b. If the price had gone from $100 to $132 in one step, the increase would have been $32/$100 = 0.32 = <u>32%</u>.

Percentage of Change: Applications p. 26

1. The area increases by <u>125%</u>.
   The old area is 10 m · 10 m = 100 m². The new area is 15 m · 15 m = 225 m². The difference is 225 − 100 = 125 m².
   To find the percent increase, we calculate the fraction 125/100 = 125%.

2. The area will decrease by <u>29.5%</u>.
   The original area is 21 cm · 29.7 cm = 623.7 cm². The new area is 17.6 cm · 25 cm = 440 cm². The difference is 623.7 − 440 = 183.7 cm². The percentage of decrease is the fraction 183.7/623.7 = 0.2945... ≈ 29.5%.

3. a. If the length of the side before scaling is x, then the length of the side after scaling is $1.15x$.
   So the side increases by $1.15x - 1.00x = 0.15x$. So as a percentage the increase is 0.15/1.00 = 15%.

   b. Let's say the side of the square measures 100 cm. After the increase, the side measures 115 cm. The area at first is 100 cm · 100 cm = 10,000 cm². After the increase, the area is 115 cm · 115 cm = 13,225 cm². The percentage increase in area is 3,225/10,000 = 0.3225 = 32.25%. Rounded to the nearest tenth of a percent, that's <u>32.3%</u>.

   You can also calculate the percentage of increase by reasoning using variables and some algebra. Let s be the side of the square before the increase. After the increase, the side is $1.15s$. The area at first is s^2, and after the increase it is $1.15^2 s^2$ or $1.3225 s^2$. The percentage increase in area is (difference in area)/(original area) $= (1.3225 s^2 - s^2)/s^2 = 0.3225 s^2/s^2 = 0.3225 = 32.25\%$. Again, that's 32.3% to the nearest tenth of a percent.

   c. No, the answer won't change. The percentage increase in area is still 32.35% whether we have a rectangle or a triangle. Whether we are calculating the area of a square, rectangle, or triangle, we multiply two dimensions: either the base and the altitude or the two sides. In all cases, the two dimensions end up being multiplied by 1.15, which means the area increases by the factor 1.15 · 1.15 = 1.3225. That scale factor corresponds to an increase of 32.25%.

4. a. Answers will vary. Please check the student's answer.
   For example: More water is used in July because it is hotter.
   b. The decreases are: For June, (215,363 − 205,849) / 215,363 = 4.0%. For July, (237,558 − 219,875) / 237,558 = 7.4%. For August, (231,777 − 204,831) / 231,777 = 11.6%.
   c. The greatest percentage decrease in water production was in <u>August</u>.
   On the graph you can see that the bar for August 2014 is less than half the length of the bar for August 2013, whereas for June and July, the bars for 2014 are more than half the corresponding bars for 2013.

5. The last increase would be less than the first. When the price went from $5.50 to $6.00, the percent increase was $0.50/$5.50 ≈ 9.1%. When it went from $6.00 to $6.50, the increase was $0.50/$6.00 ≈ 8.3%. If it goes from $6.50 to $7.00, the increase will be $0.50/$6.50 ≈ 7.7%, which is less than the first increase. Even though the actual amount of increase ($0.50) stays the same, the base price ($7.00) that we are comparing to is now larger than it was ($5.50) in the first increase. The value of the fraction 0.5/7 is smaller than the value of the fraction 0.5/5.5.

6. a. The price before tax is $69.99 / 1.07 ≈ $65.41.
   b. The percentage of increase in price with tax is ($79.99 − $69.99)/$69.99 ≈ <u>14.3%</u>.
   c. The percentage increase in price without tax is also about 14.3%.
   The new price before tax is $79.99/1.07 = $74.76. The percentage of increase is ($74.76 − $65.41)/$65.41 ≈ 14.3%.

7. a. <u>The area increases by 24.9%.</u> The original windows have the area 85 cm · 85 cm = 7,225 cm².
   The new windows will have the area 95 cm · 95 cm = 9,025 cm². The difference is 9,025 − 7,225 = 1,800 cm².
   To find the percentage of increase, calculate the fraction 1,800/7225 = 0.2491... ≈ 24.9%.

   b. Approximately 16 of the larger windows cover the same area as 20 of the smaller ones. Twenty of the smaller windows cover an area of 7,225 cm² × 20 = 144,500 cm². Dividing by the area of the larger window gives the number of larger windows needed to cover that area: 144,500 cm² / 9,025 cm² = 16.01 ≈ 16.

8. a. The $50 item has the greatest *percent* increase in price. When we compare the fractions 10/50, 10/60, and 10/70, the fraction 10/50 is the biggest because it has the smallest denominator.
   b. The price of the $70 item decreases the most in dollars. When we calculate the dollar increases, 0.12 · $50, 0.12 · $60, and 0.12 · $70, the last one is the biggest.

9. During that decade the population of the state of Kentucky increased by 3,219,000 − 3,038,000 = 181,000 people.
   As a percentage of the 1960 population, that increase is 181,000 / 3,038,000 = 0.05957... ≈ 6.0%.

Percentage of Change: Applications, cont.

10.

Year	Population	Increase for the decade	Ratio Increase / Population	% increase for the decade
1960	3,038,000	—	—	—
1970	3,219,000	181,000	0.05957...	6.0%
1980	3,661,000	442,000	0.13730...	13.7%
1990	3,685,000	24,000	0.00655...	0.7%
2000	4,042,000	357,000	0.09687...	9.7%
2010	4,340,000	298,000	0.07372...	7.4%

11. a. Which decade? From <u>1970</u> to <u>1980</u>.
   b. The slope of the line is the steepest during that decade.

12. Answers will vary. Please check the student's work.

Comparing Values Using Percentages, p. 30

1. a. difference/reference = 2 m / 6 m = 1/3 = 33.3%
   b. 180 cm / 120 cm = 3/2 = 150%
   c. 1 m / 4 m = 25%

2. Heather is 20/140 = 14.3% taller than Erica. Erica is 20/160 = 12.5% shorter than Heather.

3. a. (1) The population of Tokyo is (37.80 − 21.75)/ 21.75 = 0.7379... ≈ <u>73.8%</u> larger than the population of Delhi.
      (2) The population of Moscow is (21.60 − 15.51)/21.60 = 0.2819... ≈ <u>28.2%</u> smaller than the population of Mexico City.
   b. The difference between the populations of Seoul and New York is (25.62 − 19.95)/19.95 = 0.2842... ≈ 28.4%.
      The difference between the populations of New York and Moscow is (19.95 − 15.51)/ 15.51 = 0.2863... ≈ 28.6 %.
      So the difference between the populations of New York and Moscow is larger by <u>0.2%</u> (which is probably much less than the error in measuring the original numbers).

4. The difference in area is $\dfrac{14{,}000 - 10{,}000}{12{,}000} = \dfrac{4{,}000}{12{,}000} = 33\%$.

5. a. The percentage of difference in cost for labor is $\dfrac{\$32 - \$28}{\$30} = \dfrac{\$4}{\$30} \approx 13.3\%$.

   b. The percentage of difference in cost for labor is the same: $\dfrac{\$64 - \$56}{\$60} = \dfrac{\$8}{\$60} \approx 13.3\%$.

6. a. The shorter plant's height is 12 cm / 16 cm = 3/4 = <u>75%</u> of the taller plant's height.

   b. Write the fraction (difference)/(reference), using the shorter plant's height as the reference.
      The taller plant is 4 cm / 12 cm = 1/3 ≈ <u>33.3%</u> taller than the shorter plant.

   c. Again, write the fraction (difference)/(reference), but using the *taller* plant's height as the reference.
      The shorter plant is 4 cm / 16 cm = 1/4 = <u>25%</u> shorter than the taller plant.

   d. Again, write the fraction (difference)/(reference), but using the *average height* (14 cm) of the two plants as the reference. The average height is: ½(12 cm + 16 cm) = 14 cm. So the relative difference between their heights is 4 cm / 14 cm = 2/7 ≈ <u>28.6%</u>.

7. a. The relative height of Baby's tower is: (height of Baby's tower)/(height of Jack's tower) = 30 cm / 150 cm = 10/50 = 20/100 = 20% of the height of Jack's tower, so <u>Mary is right</u>. Elijah subtracted two heights to get a percentage; the difference in height is 120 *cm*, but that wasn't what the question asked for. Angela compared the *difference* in height instead of the height of the tower that Baby built.

   b. The school orchestra has: (difference)/(number of girls) = (26 − 14)/(14) = 12/14 = 0.85714... ≈ 85.7% ≈ 86% more boys than girls, so <u>Elijah is right</u>. Angela subtracted two counts to get a percentage; there are 12 boys more, not 12% more. Mary wrote correctly what percentage the number of girls is of the number of boys, but that wasn't what the question asked for.

Comparing Values Using Percentages, cont.

8. a. See the table on the right.

   b. You might move into County Z because of the slower increase of the crime rate, but 512 crimes is so many more than 4 that County A might well be a safer area to move into. However, it would also be good to know the total population of the areas. If there were only 4 crimes but a population of 10 in County A and 512 crimes but a population of 30 million in County Z, you might still be safer in County Z.

Number of violent crimes in 2013-2014

	County A	County Z
2013	2	454
2014	4	512
Percentage of increase	100%	12.8%

Puzzle corner:
In Ceredigion, the decrease was about 40 cases out of 93... which is nearly 1/2 or 50%.
In Conwy, the decrease is less than 30 out of 256, or barely over 10%.
In Gwynedd, the decrease is 62/200, which is 31%.
Clearly the percent of decrease was greatest in Ceredigion.

Simple Interest, p. 34

1. a. Interest: $I = P \cdot r \cdot t = \$5{,}000 \cdot 0.03 \cdot 1 = \$150$. Total to withdraw: $5,150.
   b. Interest: $I = P \cdot r \cdot t = \$3{,}500 \cdot 0.043 \cdot 4 = \$602$. Total to withdraw: $4,102.
   c. Interest: $I = P \cdot r \cdot t = \$20{,}000 \cdot 0.076 \cdot 10 = \$15{,}200$. Total to withdraw: $35,200.

2. Savings account: The interest earned is $I = P \cdot r \cdot t = \$3{,}000 \cdot 0.034 \cdot 3 = \$306$. CD:
   Since the penalty is 6 months interest, in three years Sandy would earn 2.5 year's interest.
   The total interest earned is therefore $I = P \cdot r \cdot t = \$3{,}000 \cdot 0.0392 \cdot 2.5 = \$294$.
   Sandy is better off using the savings account.

3. a. The 12.9% interest rate works out to 12.9%/12 = 1.075% a month. For a $450 tablet, with r per month and t in months (the time units have to cancel to leave only dollars), for one month that's $I = P \cdot r \cdot t = \$450 \cdot 0.01075 \cdot 1 = \$4.8375$.
   b. A day is 1/365 of a year, so each day she pays $I = P \cdot r \cdot t = \$450 \cdot 0.129 \cdot (1/365) \approx \$0.1590 = 15.9¢$.

4. Since the rate is per month, the time needs to be in months, too. So you will pay $I = P \cdot r \cdot t = \$690 \cdot 0.0109 \cdot 24 = \$180.504$ in interest.

5. Interest for the 10-month loan: $I = P \cdot r \cdot t = \$850 \cdot 0.108 \cdot 10/12 = \$76.50$.
   Interest for the 7-month loan: $I = P \cdot r \cdot t = \$850 \cdot 0.095 \cdot 7/12 = \$47.10$.
   For the shorter-term loan he would have paid $76.50 − $47.10 = $29.40 less interest.

6. Interest during the first year: $I = P \cdot r \cdot t = \$26{,}000 \cdot 0.0275 \cdot 1 = \$715$
   Interest during the last 1.5 years: $I = P \cdot r \cdot t = \$26{,}000 \cdot 0.0995 \cdot 1.5 = \$3{,}880.50$.
   The total to pay back is $715 + $3,880.5 + $26,000 = $30,595.50.

7. a. The time t is 1 year, the principal is $1,000, and the interest is $45. From the formula $I = prt$, we get $45 = \$1{,}000 \cdot r \cdot 1$ or $45 = \$1{,}000r$, from which $r = 45/1000 = 0.045 = 4.5\%$.

   b. The interest was $3,600, the time is 5 years, and the principal is $12,000. Using the formula $I = prt$, we get $3{,}600 = \$12000 \cdot r \cdot 5$ or $\$3{,}600 = \$60{,}000r$. From that, $r = 3{,}600/60{,}000 = 36/600 = 6/100 = 6\%$.

8. Let p be the original principal. In 10 years, and at 6% interest rate, that principal earns an interest of $I = prt = p \cdot 0.06 \cdot 10 = 0.6p$. In 10 years her account had $12,000, which is the original principal plus interest, or $p + 0.6p$.
   We get the equation

   $p + 0.6p = \$12{,}000$

   $1.6p = \$12{,}000$

   $p = \$12{,}000/1.6 = \underline{\$7{,}500}$

The original principal was $7,500.

Simple Interest, cont.

9. Using the formula I = *prt* we get the equation $500 = $2,000 · 0.115 · *t*. Here's the solution:

    $500 = $2,000 · 0.115 · *t*
    $500 = $230*t*
    $230*t* = $500
    t = $500/$230 = 50/23 ≈ 2.174 years

    You would have to invest it for 2.174 years or about <u>2 years 2 months</u>.

10. Using the formula I = *prt* we get the equation $350 = $1,800 · *r* · 2. Here's the solution:

    $350 = $3,600*r*
    $3,600*r* = $350
    r = $350/$3,600 = 35/360 ≈ 0.0972

    You would need an interest rate of <u>9.72%</u>.

11. The interest they paid was $4,741.48 − $4,055 = $686.48. Using the formula I = *prt*, we get:

    $686.48 = $4,055 · 0.1195 · *t*
    $686.48 = $484.5725*t*
    $484.5725*t* = $686.48
    t = $686.48/$484.5725 ≈ 1.41667 years = 1 year 5 months

You can also solve the problem this way. The interest paid was $686.48. The interest rate of 11.95% per year is very close to 1% per month, which would mean paying $40.55 per month. It would take approximately $686.48 /$40.55 ≈ 17 months to pay off the vacation package.

Puzzle corner.
a. Let's look at just one example of how compound interest is calculated. Jayden buys a $5,000 motorcycle on a credit card that has a 6% annual interest rate. Compounding the interest means that the interest is added to the principal at certain intervals, in this case each month.

Month	Monthly interest	Principal
0		$5,000
1	$5,000 · 0.005 = $25	$5,025
2	$5,025 · 0.005 = $25.125	$5,050.125
3	$5,050.125 · 0.005 ≈ $25.251	$5,075.376
4	$5,075.376 · 0.005 ≈ $25.377	$5,100.753
5	$5100.753 · 0.005 ≈ $25.504	$5,126.2567
6	$5126.257 · 0.005 ≈ $25.631	$5,151.888
7	$5151.888 · 0.005 ≈ $25.759	$5,177.647
8	$5177.647 · 0.005 ≈ $25.888	$5,203.535

b. At the end of 2 years, Jayden would pay back $5,000 · 1.005^{24} = $5,635.80.

Review, p. 40

1. a. $0.092 \cdot \$150 = \$13.80$. b. $0.458 \cdot 16$ m $= 7.328$ m. c. $0.006 \cdot 700$ mi $= 4.2$ mi.

2. a. $\$9 \cdot 0.80 = \$7.20$. New price: $\$7.20$. b. $\$6 \cdot 0.75 = \$4.50$. New price: $\$4.50$. c. $\$90 \cdot 0.70 = \$63$. New price: $\$63$.

3. $0.82p = \$23.37$
   $0.82p/0.82 = \$23.37/0.82$
   $p = \$28.50$

4. The price of the computer with tax is $\$459 \cdot 1.07 = \$491.13$.
   Andy's share of it is $\$491.13 \cdot 0.4 = \underline{\$196.45}$.
   Jack's share of it is $\$491.13 \cdot 0.6 = \underline{\$294.68}$.

5. The discount percentage is $\$27/\$180 = 3/20 = 0.15 = 15\%$.

6. a. The scale factor between the triangles is 21 cm / 7 cm = 1/3. The unknown side is therefore 12 cm/3 = 4 cm.
   The area of the larger triangle is 12 cm $\cdot$ 21 cm/2 = 126 cm^2.
   The area of the smaller triangle is 4 cm $\cdot$ 7 cm/2 = 14 cm^2.
   The area of the smaller triangle is 14/126 = 11.1% of the area of the larger triangle.

   b. If you take the ratio from the larger to the smaller triangle, the sides of the triangles are in the ratio of $\underline{3:1}$.
   Going from the smaller one to the larger one, the ratio is $\underline{1:3}$.

   c. If you take the ratio from the larger to the smaller triangle, the areas are in a ratio of $126:14 = 63:7 = \underline{9:1}$.
   Going from the smaller one to the larger one, the ratio is $\underline{1:9}$.

7. Originally, the area of the wall painting would have been 5 m $\cdot$ 3 m = 15 m^2.
   After scaling, the sides are 5 m $\cdot$ 1.2 = 6 m and 3 m $\cdot$ 1.2 = 3.6 m, so the enlarged area is 6 m $\cdot$ 3.6 m = 21.6 m^2.
   The difference in area is 21.6 m^2 − 15 m^2 = 6.6 m^2, so the percentage increase is *(difference in area)/(original area)*
   = 6.6 m^2/15 m^2 = 6.6/15 = 0.44 = 44%. An easier way to figure this is just to realize that, regardless of the actual dimensions, since the scaling of each side is 1.2, the scaling of the area is just 1.2 $\cdot$ 1.2 = 1.44, so the increase in area is 44%.

8. The difference in their times is 200 sec − 120 sec = 80 sec. Their average time was ½(200 sec + 120 sec) = 160 sec.
   a. *(Difference)/(The Old Gray Mare's time)* = 80/200 = 4/10 = 40%. Old Paint was 40% quicker than the Old Gray Mare.
   b. *(Difference)/(Old Paint's time)* = 80/120 = 2/3 = 66.7%. The Old Gray Mare was 66.7% slower than Old Paint.
   c. *(Difference)/(Average time)* = 80/160 = 1/2 = 50%. The relative difference between the two horses was 50%.

9. Assuming none of the interest was added to the principal during the time of the loan, he would owe the total amount of interest at the end of two years. For the first year, the interest is $\$4,000 \cdot 0.078 = \$312$, and for the second, it is $\$2,000 \cdot 0.078 = \$156$. The total interest is $\$312 + \$156 = \$468$.

Percent: Grade 7
Alignment to the Common Core Standards

The table below lists each lesson and next to it the relevant Common Core Standard.

Lesson	Page number	Standards
Review: Percent	13	6.RP.3 7.NS.2 7.RP.3
Solving Basic Percentage Problems	16	7.RP.3
Percent Equations	19	7.RP.3 7.EE.1 7.EE.2 7.EE.4
Circle Graphs	24	6.RP.3 7.RP.3
Percentage of Change	26	7.RP.3
Percentage of Change: Applications	29	7.RP.3
Comparing Values Using Percentages	33	7.RP.3
Simple Interest	37	7.RP.3
Review	43	7.RP.3 7.EE.4

Made in the USA
Columbia, SC
11 February 2019